Bibliographic information published by the German National Library:

The German National Library lists this publication in the National Bibliography; detailed bibliographic data are available on the Internet at http://dnb.dnb.de .

Imprint:

Copyright © 2015 GRIN Verlag
Print and binding: Books on Demand GmbH, Norderstedt Germany
ISBN: 9783668877078

This book at GRIN:

https://www.grin.com/document/454917

Daniel Häußler

Contested Worlds and a New Order After the End of the Cold War

The Scramble for Mega Sporting Event Hosting Rights as a Visualization of the New Balance of Power

GRIN Verlag

Tübingen University

Institute of Geography

"Contested Worlds" and a New Post-Cold War Order:

The Scramble for Mega Sporting Event Hosting Right as a Visualization of the Post-Cold War Order

Editor: Daniel Häußler

Tübingen, 24 August 2015

TABLE OF CONTENTS

TABLE OF CONTENTS 2

LIST OF FIGURES..3

ABSTRACT 4

1   INTRODUCTION 5

2   METHODOLOGY 6

3   THE THEMATIC FRAMEWORK..8

   3.1 POST-COLD WAR WORLD ORDER 8

   3.2 THE BACKGROUND AND MEANING OF MEGA SPORTING EVENTS........................9

   3.3 THE ASSOCIATIONS OF THE FOOTBALL WORLD CHAMPIONSHIP AND THE OLYMPIC GAMES: THE FIFA AND THE IOC 10

4   THEORETICAL PERSPECTIVE – A SUMMARY OF PETER VUJAKOVIC'S TEXT 12

5   CONNECTION BETWEEN THEORY AND PRACTICE: THE ANALYSIS OF THE SELECTION OF HOST COUNTRIES 13

   5.1 DURING COLD WAR: 1960 – 1990 14

   5.2 POST-COLD WAR: 1992 – 2022 18

   5.3 KEY FINDINGS 21

6   CONCLUSION...24

REFERENCES..25

List of Figures

Figure 1: Most popular sport in each country. Green: Football / Soccer 9

Figure 2: FIFA confederations 11

Figure 3: Venues of the Football World Cup during the Cold War............................. 14

Figure 4: Venues of the Summer Games during the Cold War 15

Figure 5: Venues of the Winter Games during the Cold War 16

Figure 6: All mega sporting events during the Cold War 17

Figure 7: Venues of the Football World Cup after the Cold War............................... 18

Figure 8: Venues of the Summer Games after the Cold War 19

Figure 9: Venues of the Winter Games after the Cold War 20

Figure 10: All mega sporting events after the Cold War 21

Figure 11: Comparison of all mega sporting events post-Cold War / during the Cold War 21

ABSTRACT

This case study addresses to the task of visualizing and describing the shifts in global politics by comparing two time periods, the Cold War and the post-Cold War era. The entirely new approach uses the awarding of mega sporting events hosting rights as an allegory for the political importance of a country: The more organizations a country is selected for, the more powerful is the relative voice on global level of politics. The most substantial argument in favor of this simplifying method is constituted in the easy and prompt access to the data due to visualizing maps. Moreover, classifying results of several countries, made by PETER VUJAKOVIC in 2005, will be compared to the interpretation of my results. With the help of this comprehensible approach you can see the topic from a different angle, and it helps to forecast the prospective development of global proportion of power. As a base, the literature research deals with an analysis of the world order during the Cold War and the shifts after the collapse of the Soviet Union. The illustration of the meaning and the background of mega sporting events completes the theoretical part. For sure, such a new methodology is susceptible to criticism, which is why I limit myself to only the central issues. Nevertheless it is a pragmatic access to a heavily manageable issue.

1 INTRODUCTION

It is well known to almost everyone to find a new (political) world order after the collapse of the Soviet Union (SU), the so-called post-Cold War order. The obvious bi-polar world went on an unclear and confused path, on which it seems to be hard to define the world's hierarchy as unipolar, bi-polar or even multi-polar. Actually we feel the competition for a better position in the global political power structure: Almost every conflict (war-like or diplomatic) is guided by the idea to improve the own position. The media serves us an extraction of these topics: The Ukraine crisis, China's land-grabbing in Africa, or even the FIFA corruption scandal. Also in the academic world, we find a huge amount of describing, analyzing and evaluating labors in different perspectives – some selected approaches will be part of the theoretical part of this case study (chapter 3.1). The central and originally injecting theoretical source constitutes the 5th chapter of MARTIN PHILIPP's book "Contested Worlds – An Introduction to Human Geography": "Nation States and Super-States: The Geopolitics of the New World Order". Its content will be shortly presented in chapter 4.

In the centerpiece of my work I want to demonstrate an entirely new approach to show the world order and its changes over the decades. By visualizing the selection of host countries of mega sporting events during the Cold War (chapter 5.1), compared to the post-Cold War world (chapter 5.2) it is provable that the global power situation has changed. With this method I will show in which way the proportion of power has shifted – relevant key-findings will be presented in chapter 5.3.

As a base, the following assumption are taken for granted: The hosting right is a matter of particular interest of every country and continent which is able to afford the protracted and cost-intensive organization. In return, the host receives a lot of attention in the global media, it shows that the country represents a dependable partner and – maybe the most important detail – it improves the population's pride to live in this region.

I want to slip in some questions which should be answered at the end of this work: Which kind of post-Cold War order do we actually find (and will be found in future) – unipolar, still bi-polar or even multi-polar (see chapter 3.2)? Preconditioned that the United States (U.S.) represents the current leader of the post-Cold War world order, which aspects could be chosen to evaluate the strength and competitiveness of potential challengers of the U.S. (see chapter 4)? How did the global proportion of power change, which countries did lose and which countries gained influence (see chapter 5.3)? As an answer and a thesis statement: Yes, global proportion of power has changed in favor of the BRICS countries (Brazil, Russia, India, China, South Africa): Today their influence is higher than during the Cold War!

2 METHODOLOGY

To visualize a shift in world order, I decided to choose the scramble for three selected mega sporting events over 62 years (1960-2022) as an indicator. The men's Football World Cup, the Summer Olympics and the Winter Olympics are the most popular mega sporting events with redistributed venues by far. I only found authoritative data about the number of viewers on a national level to strengthen this statement. However, in my opinion it isn't necessary to rank mega sporting events in a strict order of number of viewers, as the image of arranging them has a higher and more sustainable value (see chapter 1). The assumption that an assigned hosting right represents a desirable honor, as the positive aspects predominate the negative ones, can definitely took for granted. Chapter 3.2 gives some more information about the historical meaning of mega sporting events.

At first it is necessary to find out the venue of the past and also the planned events. Therefore the web sites of the particular federations enabled an easy attainable access to this fundamental information. In the next step I transferred the data into a dedicated Excel-file for a better overview (see the tables in chapter 6.1 and 6.2). To achieve the goal of a visualized result, the crucial measure was the generation of world maps, in which you can see the number of hosted mega sporting events during the particular time period: 1960-1990: Cold War and 1992-2022: post-Cold War. Thus the following maps were generated:

(1) Football World Cup (Cold War)

(2) Football World Cup (Post-Cold War)

(3) Summer Olympics (Cold War)

(4) Summer Olympics (Post-Cold War)

(5) Winter Olympics (Cold War)

(6) Winter Olympics (Post-Cold War)

(7) Total: All Mega Sporting Events (Cold War)

(8) Total: All Mega Sporting Events (Post-Cold War)

Due to the comparison of the two time periods, it is easy to show the change in global proportion of power. Mainly the maps (7) and (8) visualize this shift.

The application of this method can't be completely representative for the current world order. Some countries are more interested in arranging mega sporting events than others, e.g. due to cultural reasons, such as a particular sport being unpopular in a particular country.

Furthermore, individual economically and politically ascending countries could have a strategic plan in which the hosting right of mega sporting event is less important or even irrelevant.

My analysis starts in 1960 and contains 16, respectively 17 venues, so the events typically take place every four years. In addition the granting of the hosting right happens several years before the event, which is why the evaluation doesn't reflect exactly today's situation, but at least a tendency. Moreover the number of cases is too little to describe the shift statistically correct. Economics of scale reduce the individual costs (and thereby the hurdle for applying) by organizing two mega sporting events simultaneously – e.g. the Football World Cup and the Summer Olympics – as the stadiums built for the first event can be used again for the second event. Therefore, such an accumulation has to be viewed separately in the key-findings (chapter 5.3).

This method ignores the influences on the bulk of all countries as just a few countries are able to organize such an event. However this can't be a critique for the approach, because the conception focuses on the dominating countries of the world order.

Moreover it doesn't play a role in which way the countries have been awarded the funding for the hosting right as there are two imaginable processes: The official candidature persuaded the particular association – the FIFA or the IOC – of the good concept on the one hand, or the applicant bought the required votes from corrupted members with kickbacks on the other hand. Presumably in most cases either aspects influence the decision, but somehow or other the applying country gave proof of its strength if the candidature would be successful. Then we can take it for granted that outsiders won't find out the decisive aspects of awarding the hosting right.

During this chapter, it became clear that the applied approach is a bit vulnerable against constructive criticism but in my opinion it represents a pragmatic possibility to visualize an intricate and heavy manageable issue by using a plain access to it.

3 THE THEMATIC FRAMEWORK

To facilitate a correct embedment, the following chapter provides a theoretical frame. At first I'll present a short overview of some selected theoretical approaches which deal with the post-Cold War order and its (expected) development (chapter 3.1). Next I'll focus upon the mega sporting events: The historical meaning and importance of these events itself (chapter 3.2) will be complemented by a short introduction of the associations of the Football World Cup (FIFA) and of the Olympic Games (IOC) (chapter 3.3).

3.1 POST-COLD WAR WORLD ORDER

The rapid and unexpected changes in world politics since 1989 were to be far-reaching for the post-Cold War world order. During the conflict between East and West, the economic and political framework was quite stable, so the transformation afterwards was chaotic and unclear – even if it passed mostly peaceful – as the sense of mission and ideology got lost (CRONIN 1996, p. 238). The capitalism as the western mode of thought prevailed and according to Marx and Schumpeter it was the first truly revolutionary mode of production (ibid., p. 242). Western Europe and Japan rebuilt their economies and began to catch-up with the U.S. It is usually written that the U.S. came off as winner, while SU lost the conflict (e.g. ibid., p. 242-243; GOLDGEIER & McFAUL, p. 1-2). Thus the four decades lasting bi-polar world changed into a unipolar system: "The American victory seemed to herald Pax Americana and a new era of global stability" (MUEHLBAUER & ULBRICH 2014, p.482). Further aspects will be presented in chapter 4.

There's no doubt that hundreds and thousands of pages could be filled with discussions about the question, if U.S. came out ahead or not, and if U.S. was or is still leading the world order. Preliminary I adopt the position accepting the leading position of the U.S. as PETER VUJAKOVIC does it as well (see chapter 4). The focus of my case study is on the general shift of global political power structure and the visualization of falling and raising players and potential challengers. Therefore I would like to introduce into the mega sporting events, the meaning of the hosting right and its background.

3.2 THE BACKGROUND AND MEANING OF MEGA SPORTING EVENTS

Today, football is the most popular sport in the world (fig. 1) and even in countries with other favorites, the interest is growing. Therefore the annual pre-season trip of the most important European clubs to future merchandising markets like China or Arabia and the local enthusiasm are a good indicator. According to FIFA, about 3.2 billion people (roughly 46% of the global population) watched at least one minute of the 2010 World Cup in South Africa which reached 214 territories (KANTARSPORT 2010, p. 7). This is slightly lower than the number of people (3.6 billion in over 220 territories) who watched at least one minute of the 2012 London Olympics (IOC 2012, p. 4). Almost one billion people saw at least one minute

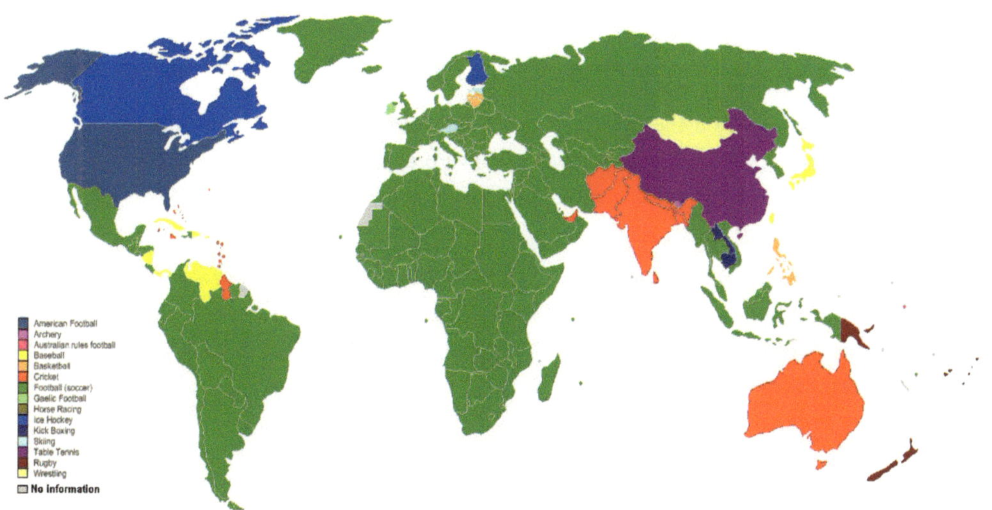

Figure 1: Most popular sport in each country. Green: Football / Soccer

Source: VOX TOPICS, confirmed by several other sources.

of the World Cup final (KANTARSPORT, p. 7) and the London Olympics' opening ceremonies (IOC 2012, p. 4). These facts underline the current meaning of mega sporting events.

In a historical view, as in a time without today's extensive broadcasting possibilities, sporting events were playing an important role, as well. Locally, it has ever been a place with great social potential (CLOSE & ASKEW & XIN 2007, p. 1). The old Romans and Greeks were used to visit spectacular sporting events, where they cheered the athletes and amused themselves – it was simply part of their culture (cf. CLOSE & ASKEW & XIN 2007, p. 44-45).

From an economical point, the framework of arranging mega sporting events has completely changed. There has been a dramatic increase in the cost of organizing the Olympics: For the 1960 Rome Summer Games, US$ 50 million was spent on "public works", while later events were a lot more expansive: 1972 Munich (US$ 850 million), 1976 Montreal (US$ 1.5 billion) and 1980 Moscow (much more than US$ 2 billion); and these additional costs couldn't be compensated in every case (CLOSE & ASKEW & XIN 2007, p. 10). The cost effectiveness doesn't have the first priority as the others aspects predominate: Causing a stir, proving its reliability and pride of the population. The scramble for mega sporting events seems to be a crystallization of the growing competition between cities, countries and continents (SHORT 2003, p. 1).

3.3 THE ASSOCIATIONS OF THE FOOTBALL WORLD CHAMPIONSHIP AND THE OLYMPIC GAMES: THE FIFA AND THE IOC

This chapter serves as a short introduction into the basic data of the 2 associations which award the hosting right of the mega sport events. Hereby I want to show the official award procedure as well. All information comes from the respective web site.

The FIFA (French: Fédération Internationale de Football Association) was founded in 1904 in Paris and is based in Zurich (Switzerland). Currently 209 national football associations have signed the membership, whereby they are authorized to participate to the qualifiers, e.g. for the World Cup (FIFA 2015a). These national associations are consolidated in 6 confederations (see fig. 2).

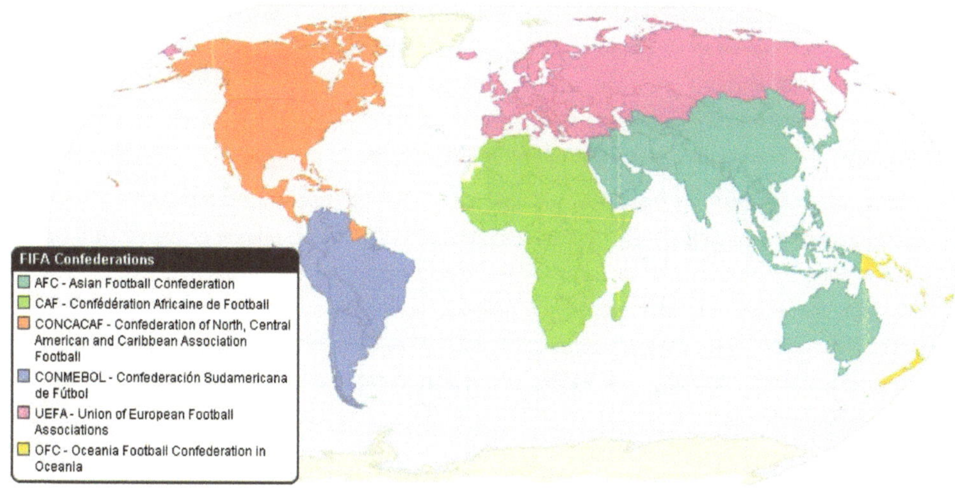

Source: CHARTSBIN.COM according to FIFA 2015c

Figure 2: FIFA confederations

The top committee is the FIFA-Congress, in which every of the 209 members has one vote – irrespective of the size of the national association. The Executive Committee consists of 25 members. These are: The President, the eight Vice-Presidents and 16 other members, who are voted by the Congress. The Executive Committee, among other things, decides about the awarding of the hosting right of the World Cup through a secret majority vote about eight to twelve years before the event (FIFA 2015b).

The FIFA inserted a rotation mechanism in 2007, after which the awarding continental association is blocked for the next two World Cups. This circumstance tampers with the explanatory power of this case study on a continental but hardly on a national level, but the mechanism ended with the award of the World Cup 2018 to Russia (FIFA, 29 Oct 2007).

Actually the media is reporting regularly on the FIFA corruption scandal. There are some notes which shall prove unfair practices concerning the awarding of the hosting right of the Football World Cup. As I mentioned in the introduction, an eye has to be kept on this issue, but it doesn't influence the explanatory power in a decisive way.

The IOC (International Olympic Committee) was founded in 1894 in Paris as well and it is headquartered in Lausanne (Switzerland). The amount of members in the IOC is limited to 115 persons. Every member has to be coopted by the Executive Board. It consists of the

President, the four Vice-Presidents and ten separately voted members. The proximate admission of the coopted person has to be confirmed by the IOC-Session. The IOC-Session's task includes also the selection of primary possible venues for the Olympic Games, for which every member has one vote. After that, it comes to a decision by a secret majority vote through the Executive Committee (IOC 2015a).

Thus the proceedings of the FIFA and the IOC have the same basic framework and it became clear that a candidature for the Football World Cup or the Summer or Winter Games has to take a lot of hurdles to achieve the hosting right. The deciding institution put the concept, the economic, socio-economic, political, ecological and financial frame and the reliability through their paces and agrees just if it is absolutely sure of it. It can be noted that only several mighty countries are eligible and willing to organize a mega sporting event (cf. SHORT 2003, p. 1-2).

4 THEORETICAL PERSPECTIVE – A SUMMARY OF PETER VUJAKOVIC'S TEXT

MARTIN PHILIPPS edited "Contested Worlds – An Introduction to Human Geography" in 2005, whereof chapter 5 ("Nation States and Super-States: The Geopolitics of the New World Order", written by PETER VUJAKOVIC) will be shortly summarized. As the main thesis, the chapter declares the U.S. as the undisputed winner of the Cold War and consequently as the nation which is challenged by the others. According to the author, this scramble has distinctly intensified since the end of Cold War (VUJAKOVIC 2005, p. 166-167). At this point I want to add the question, if the continental or national origin of challengers could have been changed over the decades. Chapter 5 of this case study will give you an informative answer on that by analyzing the selection of host countries.

Preliminary VUJAKOVIC points out the definitive aspects of a state (partitioned space or territory, controlled by one sovereign power) and a nation (state, filled up with people, culture and community) (ibid. p. 153-154) and supplements the definitions of a Nation State (a nation with an ethnically homogenous population, e.g. Korea) and a super-state (world's contest leading country or supranational union, e.g. U.S.) (ibid. p. 161-162, 165). This content isn't that relevant for this case study, but it shouldn't remain unmentioned.

The author discusses shortly – but in a substantive way – with the geographical concept of territoriality which is used to justify and to evaluate the strength and competitiveness of potential challengers of the U.S. (ibid. p. 156-160). The following results are presented, which are going to be verified in my analysis in chapter 5.

China could definitely be a serious challenger, if their economical rate of growth goes on. At this point we have to keep in mind that the book was published in 2005, but the economy of the "Middle Kingdom" is still growing fast. In contrast, **Japan** is described as a "post-hegemonic power" (ibid. p. 171) and it seems to be no more than a regional power in the Pacific Region. **Russia** is still profiting by its useful natural resources, but stands at "its crossroads" (ibid. p. 172) between a regional power within Europe and Asia and a global player alongside the U.S. The **European Union** is the sole player that is increasing its territory in a mostly legitimized way, while **Germany** is a draft horse and peace-keeping protagonist for Europe but it won't influence the world order in a strongly way (ibid. p. 169-178).

As a next step, the following analysis visualizes the shift in global distribution of power, checks the classification of PETER VUJAKOVIC and tries to forecast the prospective development. Connection between Theory and Practice: The Analysis of the Selection of Host Countries

This part of the case study will connect the aforesaid theory about the competition for a better position in the world order with the new approach of visualizing the scramble for mega sporting events. Each subchapter (5.1 During Cold War and 5.2 Post-Cold War) contains four maps: Football World Cup, Summer Games, Winter Games and an all-including overview. The maps were generated by STEPMAP.DE, a free accessible internet service. It has an advantage over a professional GIS, as the availability of map-data and the handling was much easier. Pertained countries are colored: Green Color stands for one arrangement, Orange Color for two arrangements and Red Color for three arrangements. I will describe each map shortly and add some information as the case may be, but I will concentrate on the concluding comparison. Thereby I can formulate the key-findings, which give you an answer to the prior questions.

4.1 DURING COLD WAR: 1960 – 1990

Figure 3: Venues of the Football World Cup during the Cold War

Source: Own creation with stepmap.de according to FIFA 2015d

As you can see in fig. 3, the awarding of the Football World Cup was concentrated on Europe, South America and Mexico. Four of eight events were awarded to Europe, and between 1966 and 1990 it was even assigned every other Football World Cup.

Fig. 4 shows you the venues of the Summer Games during the Cold War. All events have taken place on the northern hemisphere and no country has been assigned two times.

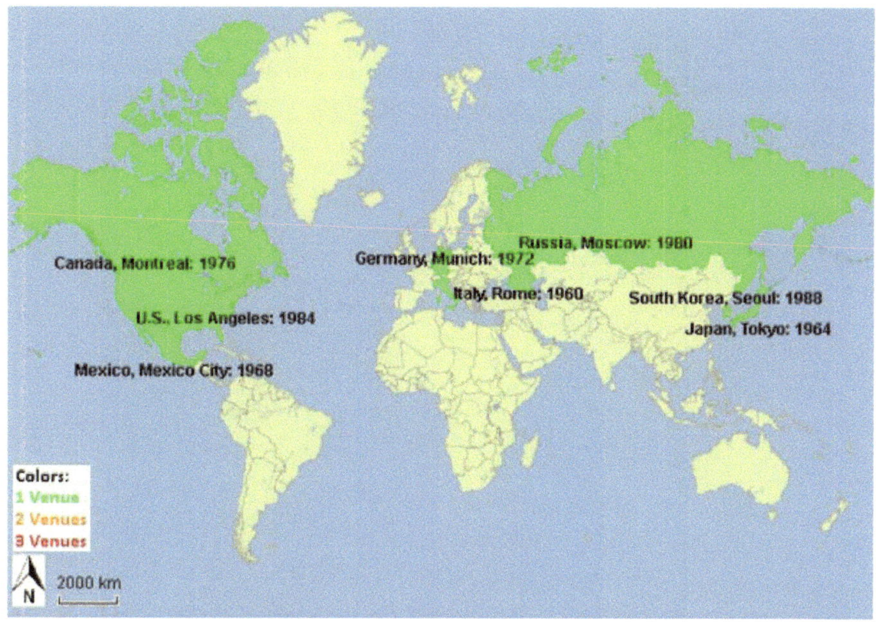

Figure 4: Venues of the Summer Games during the Cold War

Source: Own creation with stepmap.de according to IOC 2015b

Thirdly, fig. 5 shows the Winter Games. The information of this map has to be taken with a pinch of salt due to geographical requirements to organize Winter Games: It is absolutely necessary to guarantee a reliable snow coverage, which is why many countries wouldn't be able to submit a promising candidature.

Moreover Innsbruck, Austria was awarded in 1964 and once again in 1976. Originally Denver, U.S. was chosen to organize the Winter Games in 1976, but Innsbruck replaced Denver due to a public participation because the capital investment in the preparation of the Games was seen as a waste. Innsbruck organized wonderful Winter Games in 1964 and already had the needed infrastructure, so it prevailed in a second ballot. (DENVER PUBLIC LIBRARY 2015). All Winter Games took place in the northern hemisphere, more precisely in the classical triad Northern America, Europe and Japan.

Figure 5: Venues of the Winter Games during the Cold War

Source: Own creation with stepmap.de according to IOC 2015b

The all-including overview (fig. 6) clarifies the impression, which already should be arisen: Most of the mega sporting events were organized by the highly developed countries of the triad. U.S. and Mexico were awarded three times, Southern America only was chosen for two Football World Cups while Africa, Central Asia, Australia and Oceania were completely ignored. The explanatory effect will be described more rich in content by comparing this result with the post-Cold War era.

Figure 6: All mega sporting events during the Cold War

Source: Own creation with stepmap.de according to FIFA 2015d & IOC 2015b

4.2 POST-COLD WAR: 1992 – 2022

Suddenly we are now spotting new hosts on the next map, which visualizes the proportion of the Football World Cup from 1990 to 2022 (fig. 7). Previously unstated countries were selected to arrange this mega sporting event in 2010 (South Africa), 2014 (Brazil) and 2022 (Qatar). The other tournaments are apportioned among well-known countries like Germany, France, Russia, Japan and South Korea and of course U.S. Every continent – except Australia including Oceania – was considered at least once and no country was chosen two times.

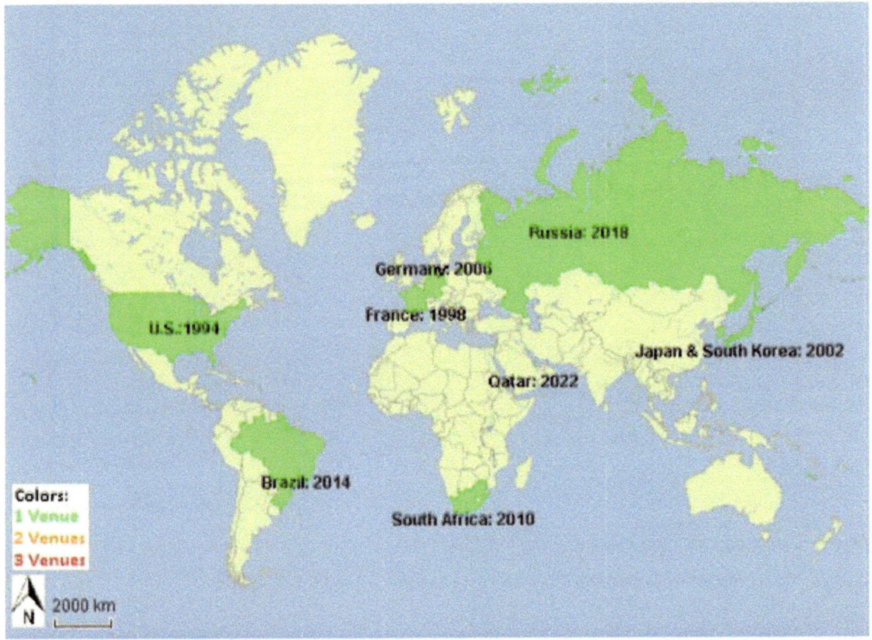

Figure 7: Venues of the Football World Cup after the Cold War

Source: Own creation with stepmap.de according to FIFA 2015d

Fig. 8 shows the venues of the past and planned Summer Games after the Cold War. Every continent – except Africa – was selected at least once to arrange this mega sporting event and no country was selected two times or even more often. Unsurprisingly the Summer Games in 1992 were characterized by a record participation due to the political change: No country boycotted the Games, while the prior SU participated under the flag of the Commonwealth of Independent States (CIS) (KLUGE 2000, p. 349-351).

It was generally expected, that Athens will arrange the Summer Games in 1996 due to the centenarian jubilee of the modern Olympic Games, but Atlanta prevailed in the last round of ballot. It was supposed, that Coca Cola, one of the most important sponsors and domiciled in Atlanta, influenced the decision (KLUGE 2000, p. 677).

In 2016 the Olympic Games in Rio de Janeiro, Brazil will take place in Southern America for the first time in Olympic history.

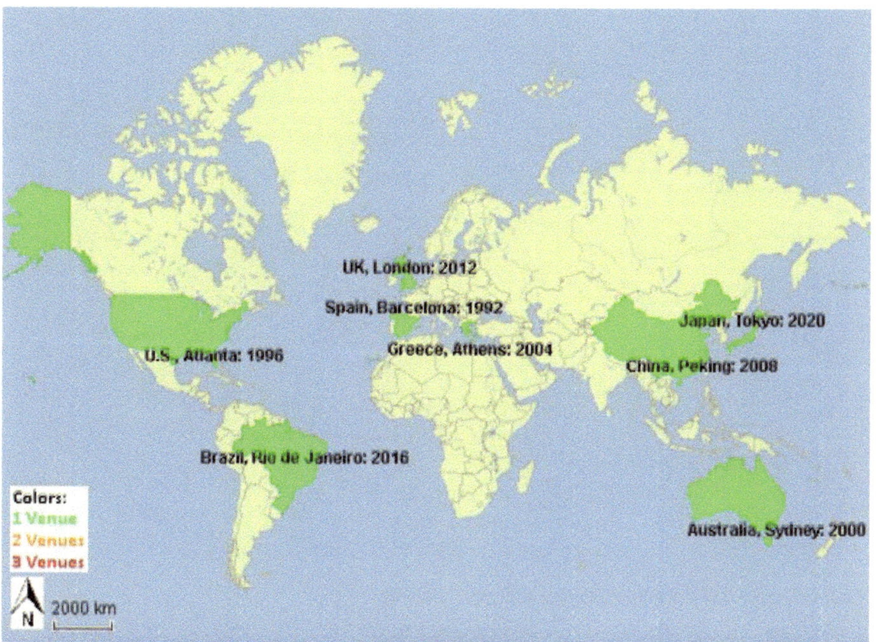

Figure 8: Venues of the Summer Games after the Cold War

Source: Own creation with stepmap.de according to IOC 2015b

As I mentioned before, the subsumption of the map, which shows the venues of the Winter Games from 1992 to 2022 (fig. 9), should be considered carefully due to climate and geographical aspects. Nevertheless, it is remarkable, that the Winter Games happened again just in the Northern hemisphere. Northern America was selected for 2 Winter Games, Europe for 3 Winter Games and Asia even for 4 Winter Games – preconditioned, that Sotschi,

Russia is counted among Asia – and 3 of them in a row: 2014 (Sotschi, Russia), 2018 (Pyongyang, South Korea) and 2022 (Peking, China). The decision in favor of Peking was made on the 31st of July 2015 in Kuala Lumpur, Malaysia, so this information is quite up-to-date (IOC 2015b).

The Winter Games took place in 1992 and again in 1994 due to the idea that the Summer Games and the Winter Games are alternating every two years.

Figure 9: Venues of the Winter Games after the Cold War

Source: Own creation with stepmap.de according to IOC 2015b

Fig. 10 summarizes the information of all mega sporting events after the Cold War: Every continent was selected at least once, but Africa and Australia just one time.

Many European countries hosted at least one event: Norway, UK, Germany, Italy, France, Spain and Greece. It is also conspicuous, that a large area of the Asian part of the map is orange-colored, so Russia, China and also South Korea arranged two mega sporting events.

Japan and U.S. are the single countries with 3 hosting rights.

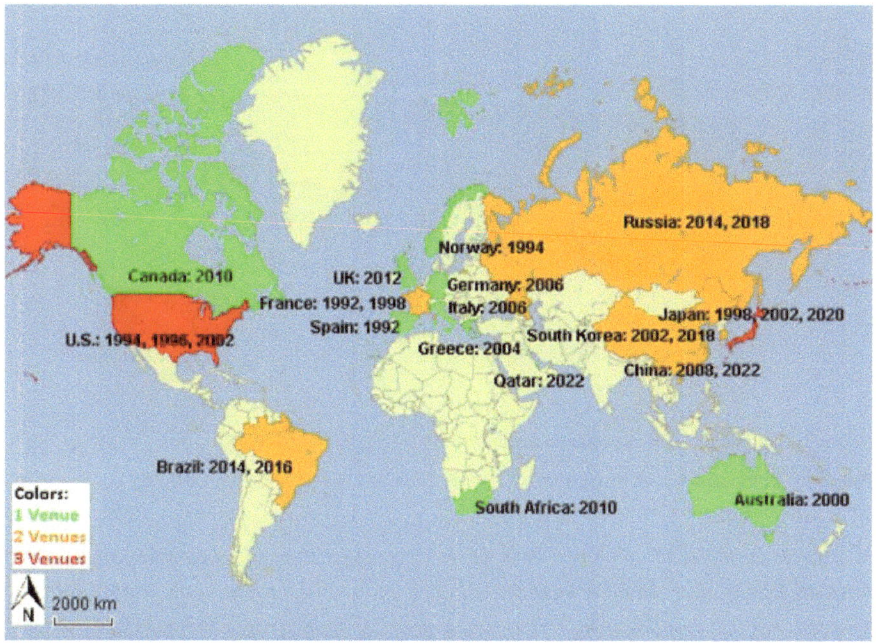

Figure 10: All mega sporting events after the Cold War

Source: Own creation with stepmap.de according to FIFA 2015d & IOC 2015b

4.3 KEY FINDINGS

This chapter includes two parts: First, the essential results and interpretations will be summarized. Second I will refer to and tie into the main aspects and evaluations of PETER VUJAKOVIC, which were presented briefly in chapter 4.

The eight maps in the chapters 5.1 and 5.2 only visualize the information about the proportion of the Football World Cup, the Olympic Summer Games and the Olympic Winter Games. In the next step, I will show which shifts in the world order could be determined. For this fig. 11 gives a good overview. It shows the sum of all mega sporting events during the Cold War (cf. fig. 6 in chapter 5.1) and after the Cold War (cf. fig. 10 in chapter 5.2). It isn't important, that the lettering is too small as you should focus on the coloring.

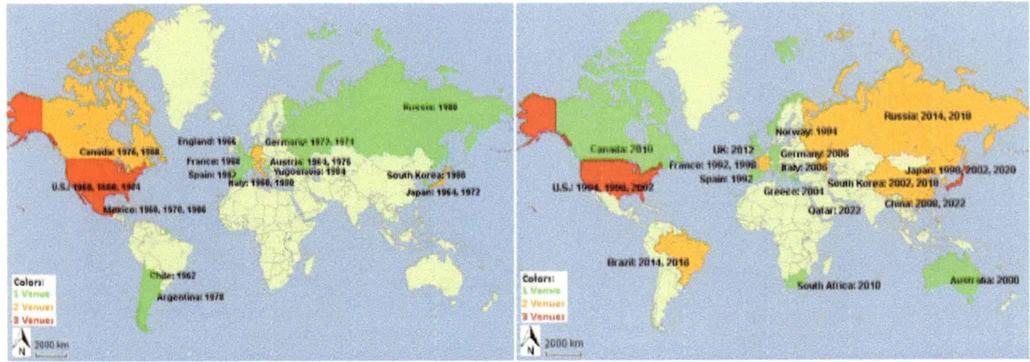

Source: Own creation with stepmap.de according to FIFA 2015d & IOC 2015b

Figure 11: Comparison of all mega sporting events post-Cold War / during the Cold War

In general, a whale of difference isn't visible: Especially the triad, consisting of Northern America, Europe and Japan, is well colored, while Southern America, Africa, Middle East and Southern Asia stay largely white on the maps, even if there is a slight and selective increase. The U.S. represent a red area on both maps, which means that they've arranged 3 mega sporting events in each time period. Generally you can see, that Asia developed from a green region to an orange and red one due to an excess of hosting rights. It's also remarkable, that, as I mentioned at the beginning of chapter 5.2, on the right hand side, you can see more different involved countries, though the compared time periods are isochronous.

On a national level, I want to broach some individual shifts. Mexico, which had 3 hosting rights during the Cold War, had its last appearance on the global sporting scene in 1986. Going by the implications of this study, this country lost a lot of influence in global politics. This statement can't be proven exactly, as a meaningful analysis of Mexico's economy and politics would go beyond the scope of this work. In chapter 1, I asked the change of the global proportion of power, and I anticipated it to have changed in favor of the BRICS countries. At this point this needs to be clarified.

During the Cold War, Brazil (0), Russia (1), India (0), China (0) and South Africa (0) only arranged one mega sporting event, the Olympic Summer Games in 1980 in Moscow. In the post-Cold War time period, Brazil (1), Russia (2), India (0), China (2) and South Africa (1) arranged even six mega sporting events. India seems to be an exception, as Cricket is much more popular than Football in this country (cf. fig. 1). Currently India is trying to apply for the Summer Games in 2024, even if the president of the IOC, Thomas Bach, evaluated it as too early (e.g. HANDELSBLATT 2015). In this angle of view, the BRICS countries could play a major role in the future, especially because interdependence could be guarantee for world-wide peace.

In 2005, PETER VUJAKOVIC forecasted that **China** could definitely be a serious challenger. Based on my approach, this evaluation can be confirmed as it has an increasing number of hosting rights of mega sporting events. According to him, **Japan** is a "post-hegemonic power" (VUJAKOVIC 2005, p. 171) with an alleviative influence on global politics, while my analysis is showing the opposite: Japan had the hosting right for both Olympic Games during the Cold War, will have arranged both Games in the post-Cold War period, and even added the Football World Cup in 2002. VUJAKOVIC says, that **Russia** stands at "its crossroads" (ibid. 172), but according to me, the current development of the position of Russia indicates an increasing influence after the collapse of the SU as the number of hosting rights increased in only a short period of time. In contrast, the **European Union** will have arranged as many mega sporting events in the post-Cold War period as during the Cold War, though VUJAKOVIC evaluated it as the sole player that is increasing its territory and influence in a mostly legitimized way. **Germany** itself will have arranged only one mega sporting event in the post-Cold War period, namely the Football World Cup in 2006, while it hosted the Summer Games in 1972 and the Football World Cup in 1974 during the Cold War, though Hamburg is preparing a candidature for 2024 or 2028 (DOSB 2015). I would accord with VUJAKOVIC, who assumes Germany as a draft horse and peace-keeping protagonist for Europe, but won't influence the world order in a strong way (ibid. p. 169-178). Moreover, his assumption, that the **U.S.** never lost the leading position in global politics can be confirmed at least partly, as they arranged a lot of mega sporting events, which is why this country is red-colored on both maps. But keep in mind that the last mega sporting event happened in 2002, and Asia is getting more and more hosting rights.

Another key-finding must be a slight self-criticism, as the global proportion of power can't be expressed that precisely in such a simple way. I mentioned some points on this in chapter 2, but it shouldn't derogate the additional value of this case study.

5 CONCLUSION

The topic „The Scramble for Mega Sporting Event Hosting Right as a Visualization of the post-Cold War Order" could be presented on a much larger scale. First it could be reflected on the number of candidatures, as well as the effort of a country, irrespective of the number of awarded contracts. Moreover would it be more representative to expand the approach on other kinds of sporting events, e.g. handball, Formula One, swimming or even chess. From my point of view, sports like ice hockey and rugby would be less eligible due to the regional differences concerning the popularity. So another challenge for new far-reaching work would be the selection of the sport. Furthermore, the description of the results is a matter of interpretation – a typical issue in Political Geography and Human Geography in general.

A study about the role of the media would upgrade the analysis as well. It would be helpful to ascertain, how the favored countries exploit the awarding of a hosting right, and especially if there are differences between individual countries, e.g. U.S., Russia and China.

However, this approach definitely helps to understand the shifts in the world, especially by the comparison of the Cold War and the post-Cold War era. The essential results are:

- Today there are more players on global politics than during the Cold War.
- The U.S. is still leading the global politics, but in the long term they have to be wary of serious challengers.
- Currently you can witness a shift in favor of Asian countries whereof China is jutting.
- The BRICS countries are also among the aspiring countries.
- Most of the South American, African and Oceanian countries don't play a major role in global politics – neither in the past nor in the future.
- The classification of PETER VUJAKOVIC can be extensively confirmed.

It should be clear that a bid for a mega sporting event implicates more than economic reasoning. Against the background of a shifting proportion of global power, you see the circumstances form a different angle. So the population of an elected country can be proud to live in such a well-off country.

At last, I would like to give you a continuative reading tip. MAXIMILIAN TERHALLE recently published "The Transition of global Order. Legitimacy and Contestation", wherein the differences between U.S. and China are deconstructed. This great power rivalry is described without getting lost in Liberalism's reassuring radiance or Realism's despairing darkness.

REFERENCES

CHARTSBIN.COM (2015): Map of the World with the six Confederations. On the Internet: http://chartsbin.com/view/owq#disqus_thread (Retrieved 19 August 2015). Confirmed by FIFA 2015c.

CLOSE, P.; ASKEW, D.; XIN, X. (2007): The Beijing Olympiad. The Political Economy of a Sporting Mega-Event. New York.

CRONIN, J. E. (1996): The World the Cold War made. Order, Chaos and the Return of History. New York and London.

DENVER PUBLIC LIBRARY (2015): The Denver that never was: 1976 Winter Olympic Games. On the Internet: https://history.denverlibrary.org/news/denver-never-was-1976-winter-olympic-games (Retrieved 19 August 2015).

DOSB – DEUTSCHER OLYMPISCHER SPORTBUND E.V. (2015): Olympiabewerbung Hamburg 2024/2028. On the internet: http://www.dosb.de/de/olympia/olympiabewerbung/ (Retrieved 19 August 2015).

FIFA – FÉDÉRATION INTERNATIONALE DE FOOTBALL ASSOCIATION (29 Oct 2007): Rotation ends in 2018. On the internet: http://www.fifa.com/worldcup/news/y=2007/m=10/news=rotation-ends- 2018-625122.html (Retrieved 19 August 2015).

FIFA – FÉDÉRATION INTERNATIONALE DE FOOTBALL ASSOCIATION (2015a): Who we are. On the internet: http://de.fifa.com/about-fifa/who-we-are/index.html (Retrieved 19 August 2015).

FIFA – FÉDÉRATION INTERNATIONALE DE FOOTBALL ASSOCIATION (2015b): Committees. On the internet: http://de.fifa.com/about-fifa/committees/index.html (Retrieved 19 August 2015).

FIFA – FÉDÉRATION INTERNATIONALE DE FOOTBALL ASSOCIATION (2015c): Associations. On the internet: http://www.fifa.com/associations/index.html (Retrieved 19 August 2015).

FIFA – FÉDÉRATION INTERNATIONALE DE FOOTBALL ASSOCIATION (2015d): World Cup Timeline. On the internet: http://www.fifa.com/worldcup/ (Retrieved 19 August 2015).

GOLDGEIER, J. M.; MCFAUL, M. (2003): Power and Purpose. U.S. Policy toward Russia after the Cold War. Washington.

Handelsblatt (2015): Bach rät Indien von Bewerbung für Olympia 2024 ab. On the internet: http://www.handelsblatt.com/olympia-2024-bach-raet-indien-von-bewerbung-fuer-olympia-2024-ab/11699592.html (Retrieved 19 August 2015).

IOC – International Olympic Committee (2012): London 2012 Olympic Games. Global Broadcast Report. On the Internet: http://www.olympic.org/Documents/ IOC_Marketing/Broadcasting/London_2012_Global_%20Broadcast_Report.pdf (Retrieved 19 August 2015).

IOC – International Olympic Committee (2015a): The Organisation. On the internet: http://www.olympic.org/about-ioc-institution?tab=organisation (Retrieved 19 August 2015).

IOC – International Olympic Committee (2015b): Olympic Games. On the internet: http://www.olympic.org/olympic-games (Retrieved 19 August 2015).

KANTARSPORT (2010): 2010 FIFA World Cup South Africa. Television Audience Report. On The Internet: http://www.fifa.com/mm/document/affederation/tv/01/47/32/73/2010fifa worldcupsouthafricatvaudiencereport.pdf (Retrieved 19 August 2015).

KLUGE, V. (2000): Olympische Sommerspiele. Die Chronik Teil 4. Berlin.

MUEHLBAUER, M. S.; ULBRICH, D. J. (2014): Ways of War. American Military from the Colonial Era to the Twenty-First Century. New York & Abingdon.

SHORT, J. (2003): Going for Gold: Globalizing the Olympics, Localizing the Games. In: Global Metropolitan: Globalizing Cities in a Capitalist World, p. 86-108. London.

TERHALLE, M. (2015): The Transition of Global Order. Legitimacy and Contestation. New York.

VOX TOPICS (2015): The most popular sport in every country. On the Internet: http://www.vox.com/2014/7/3/5868115/most-popular-sports-world-cup (Retrieved 19 August 2015).

VUJAKOVIC, P. (2005): Nation States and Super-States: The Geopolitics of the New World Order. In: PHILIPPS, M. (Ed.) (2005): Contested Worlds. An Introduction to Human Geography.

YOUR KNOWLEDGE HAS VALUE

- We will publish your bachelor's and
  master's thesis, essays and papers

- Your own eBook and book -
  sold worldwide in all relevant shops

- Earn money with each sale

Upload your text at www.GRIN.com
and publish for free